ANIMALS OF THE GRASSLANDS

AWESOME OSTRICHES

THERESA EMMINIZER

PowerKiDS press.
New York

Published in 2021 by The Rosen Publishing Group, Inc.
29 East 21st Street, New York, NY 10010

Portions of this work were originally authored by Therese Harasymiw and published as *Ostriches*. All new material in this edition authored by Theresa Emminizer.

Editor: Theresa Emminizer
Book Design: Michael Flynn

Photo Credits: Cover, p. 1 Jurie Maree/Shutterstock.com; series background Chantal de Bruijne/Shutterstock.com; p. 5 semue85/Shutterstock.com; p. 7 Dr Ajay Kumar Singh/Shutterstock.com; p. 9 Peter Titmuss/Shutterstock.com; p. 11 Ondacaracola/Shutterstock.com; p. 13 EcoPrint/Shutterstock.com; p. 15 Yogiee/Shutterstock.com; p. 17 Ayca Yalciner/Shutterstock.com; p. 19 Jonathan Pledger/Shutterstock.com; p. 21 wayak/Shutterstock.com; p. 22 Aaron Amat/Shutterstock.com.

Library of Congress Cataloging-in-Publication Data

Names: Emminizer, Theresa, author.
Title: Awesome ostriches / Theresa Emminizer.
Description: New York : PowerKids Press, [2021] | Series: Animals of the grasslands | Includes index.
Identifiers: LCCN 2019048268 | ISBN 9781725316867 (paperback) | ISBN 9781725316881 (library binding) | ISBN 9781725316874 (6 pack)
Subjects: LCSH: Ostriches—Juvenile literature.
Classification: LCC QL696.S9 E457 2021 | DDC 598.5/24—dc23
LC record available at https://lccn.loc.gov/2019048268

Manufactured in the United States of America

CPSIA Compliance Information: Batch #CSPK20. For Further Information contact Rosen Publishing, New York, New York at 1-800-237-9932.

CONTENTS

The Biggest Bird

Ostriches are the biggest birds in the world. They can be up to 9 feet (2.7 m) tall. They can weigh as much as 350 pounds (158.7 kg). Ostriches can be found across Africa. They live in **grasslands** and deserts.

Short Wings

Ostriches have short wings. They can't fly because their bodies are too heavy. Instead, they use their wings to help them **balance**. As ostriches run, they hold their wings out to the side. This helps them keep from falling down.

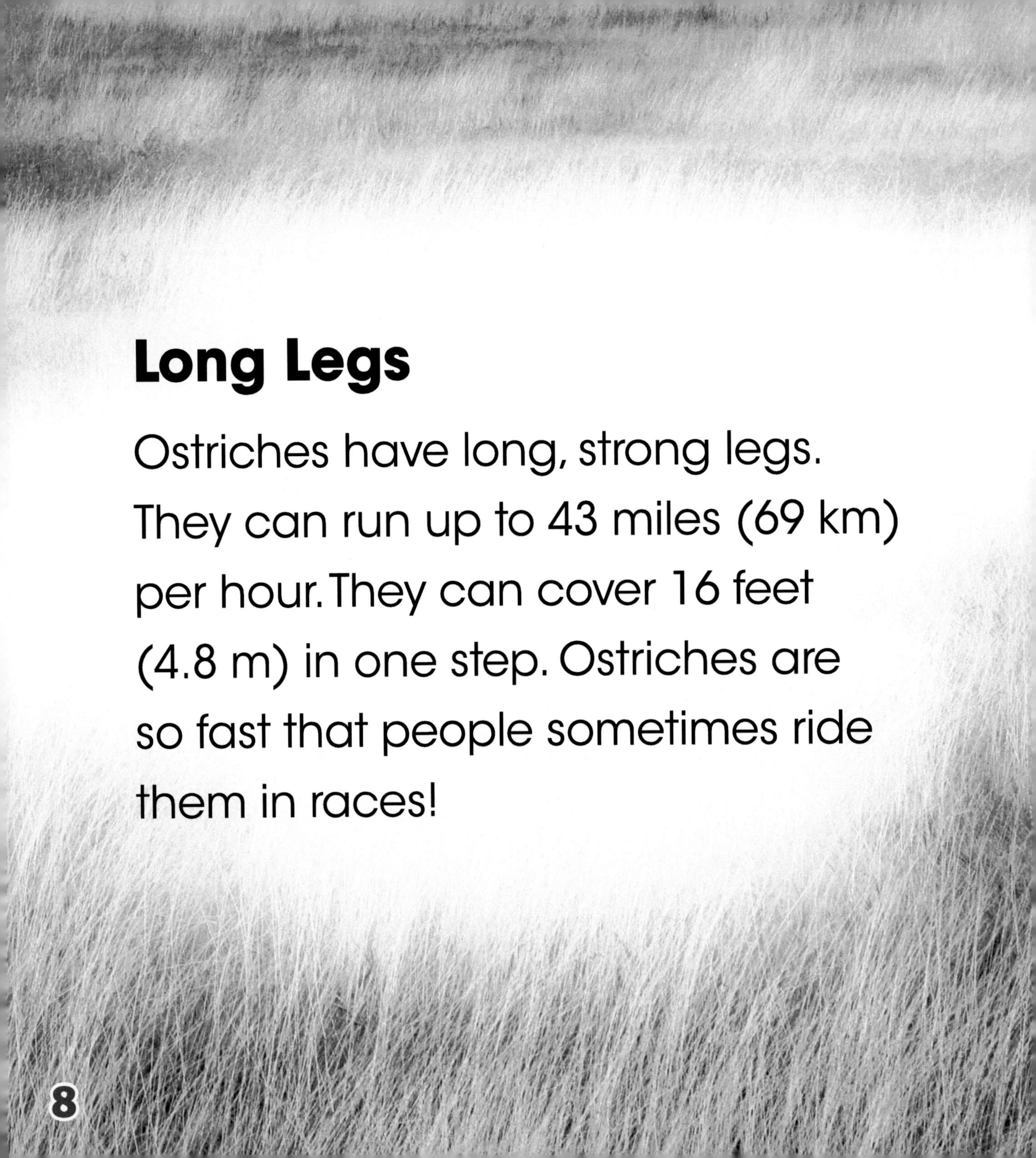

Long Legs

Ostriches have long, strong legs. They can run up to 43 miles (69 km) per hour. They can cover 16 feet (4.8 m) in one step. Ostriches are so fast that people sometimes ride them in races!

Wild Weapons

Ostriches have very big eyes to keep a lookout for **predators**. If **attacked**, ostriches use their legs to keep themselves safe. An ostrich's kick is strong enough to kill a lion! Ostriches have one sharp claw on each of their two-toed feet.

Males and Females

Female, or girl, ostriches are grayish brown in color. Male, or boy, ostriches have black and white feathers. Male ostriches do a special dance to show off for females. They shake their wings and feathers and wag their head.

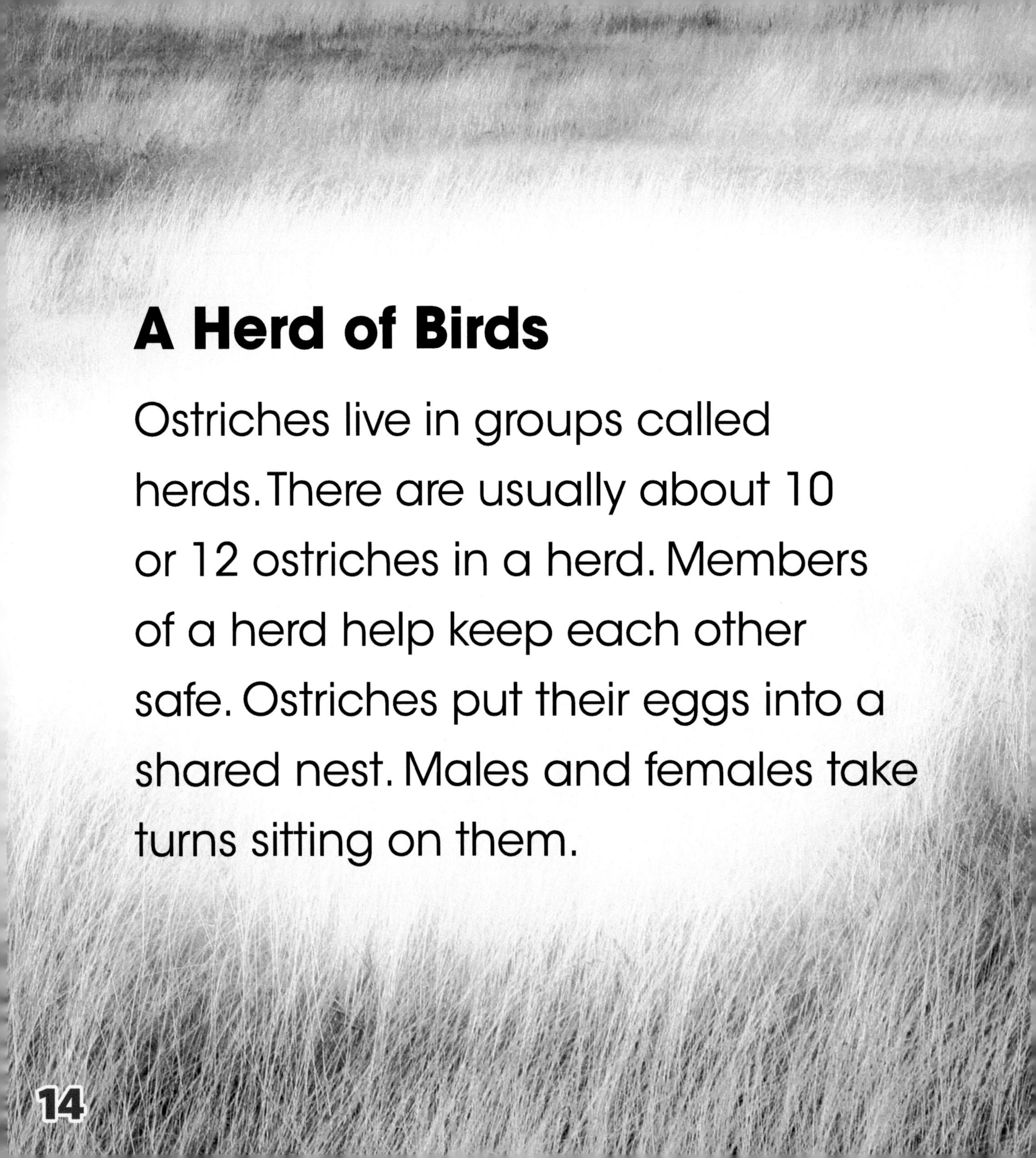

A Herd of Birds

Ostriches live in groups called herds. There are usually about 10 or 12 ostriches in a herd. Members of a herd help keep each other safe. Ostriches put their eggs into a shared nest. Males and females take turns sitting on them.

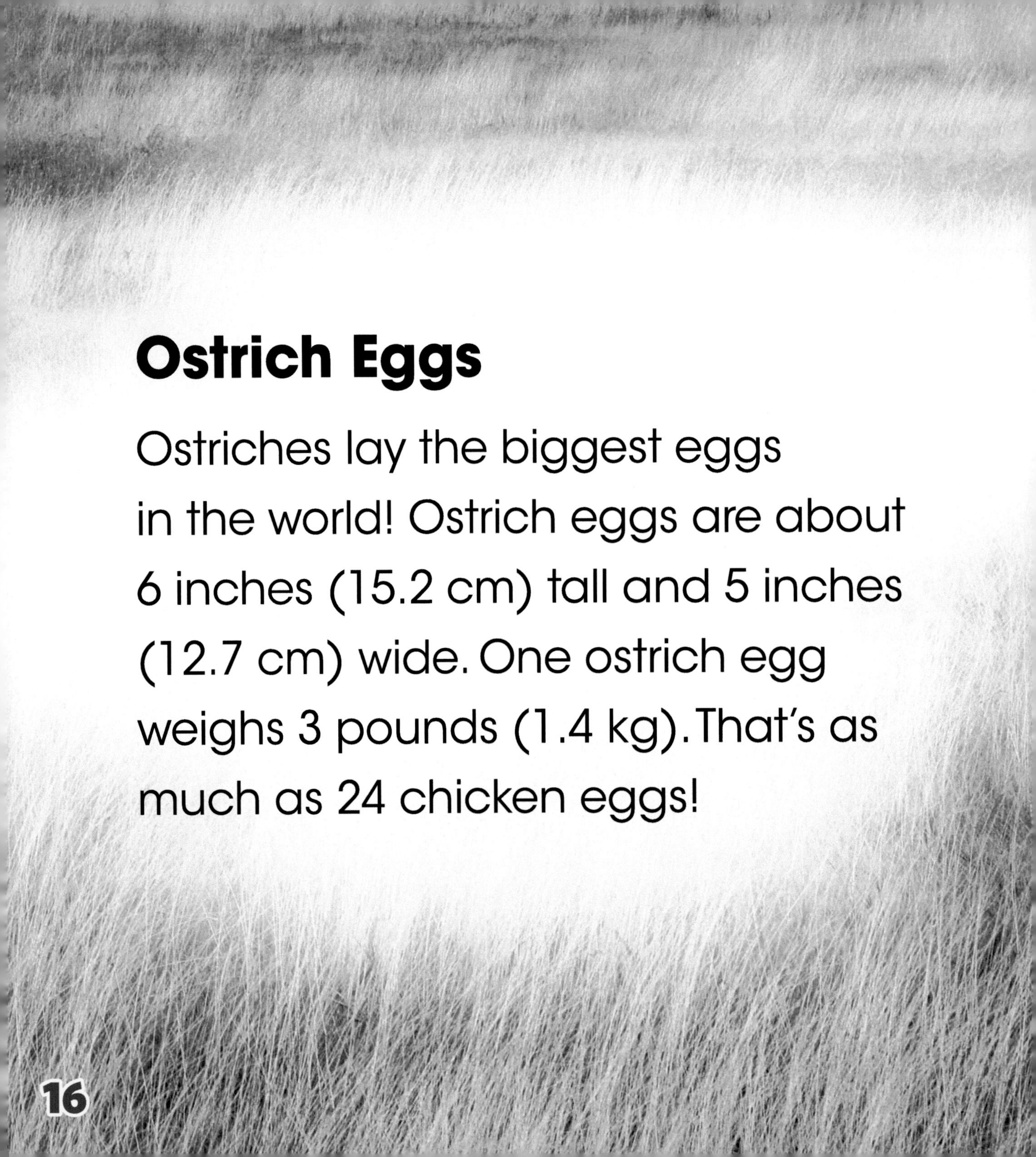

Ostrich Eggs

Ostriches lay the biggest eggs in the world! Ostrich eggs are about 6 inches (15.2 cm) tall and 5 inches (12.7 cm) wide. One ostrich egg weighs 3 pounds (1.4 kg). That's as much as 24 chicken eggs!

ostrich egg
chicken egg

What Ostriches Eat

Ostriches are omnivores. That means they eat both plants and meat. They eat roots, leaves, seeds, bugs, snakes, and lizards. Ostriches don't have teeth. They eat sand and stones to help break up food in their stomach.

Special Adaptations

Ostriches have **adapted** to live in grasslands. They can go for days without drinking water. Their grayish brown feathers **blend** in with the soil. Their big eyes can spot danger from far away. Their long legs and claws help them fight predators.

Fancy Features

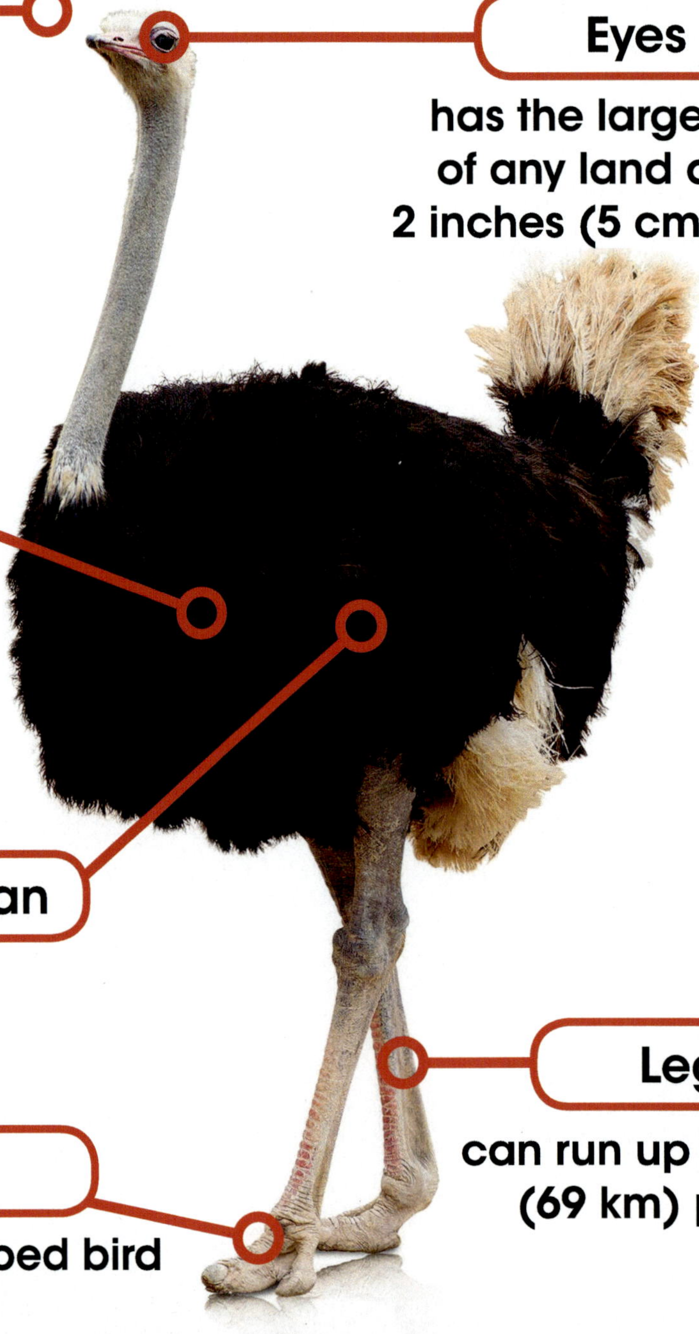

Height

up to 9 feet
(2.7 m) tall

Eyes

has the largest eyes
of any land animal,
2 inches (5 cm) around

Weight

up to 350 pounds
(158 kg)

Average Life Span

up to 40 years
in the wild

Legs

can run up to 43 miles
(69 km) per hour

Feet

the only two-toed bird

GLOSSARY

adapt: To change to suit conditions.

attacked: Harmed by something.

balance: To keep something steady so it doesn't fall.

blend: To mix in with something.

grassland: Land on which grass is the main plant life.

predator: An animal that hunts other animals for food.

INDEX

WEBSITES

Due to the changing nature of Internet links, PowerKids Press has developed an online list of websites related to the subject of this book. This site is updated regularly. Please use this link to access the list: www.powerkidslinks.com/aotg/ostriches